Henry Wadsworth Longfellow

The Hanging of the Crane

Henry Wadsworth Longfellow

The Hanging of the Crane

ISBN/EAN: 9783337105358

Printed in Europe, USA, Canada, Australia, Japan

Cover: Foto ©berggeist007 / pixelio.de

More available books at **www.hansebooks.com**

THE

HANGING OF THE CRANE

BY

HENRY WADSWORTH LONGFELLOW

WITH ILLUSTRATIONS

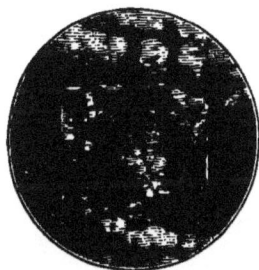

BOSTON
JAMES R. OSGOOD AND COMPANY
Late Ticknor & Fields, and Fields, Osgood, & Co.
1875

WELCH,
BIGELOW,
& CO.
UNIVERSITY
PRESS.

LIST OF ILLUSTRATIONS.

The Engravings by A. V. S. ANTHONY, *under whose supervision the book is prepared, and* W. J. LINTON.

[The Emblematical and Ornamental Vignettes are drawn by JOHN J. HARLEY.]

THE HANGING OF THE CRANE

I.

THE lights are out, and gone are all the guests

That thronging came with merriment and jests

   To celebrate the Hanging of the Crane

In the new house, — into the night are gone;

But still the fire upon the hearth burns on,

   And I alone remain.

O fortunate, O happy day,

When a new household finds its place

Among the myriad homes of earth,

Like a new star just sprung to birth,

And rolled on its harmonious way

Into the boundless realms of space!

So said the guests in speech and song,

As in the chimney, burning bright,

We hung the iron crane to-night,

And merry was the feast and long.

II.

AND now I sit and muse on what may be,

And in my vision see, or seem to see,

Through floating vapors interfused with light,

Shapes indeterminate, that gleam and fade,

As shadows passing into deeper shade

Sink and elude the sight.

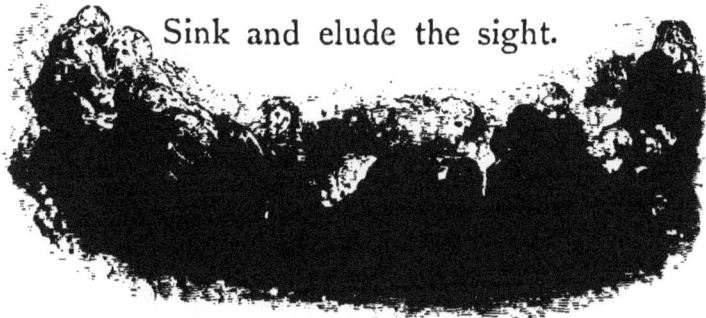

For two alone, there in the hall,

Is spread the table round and small;

Upon the polished silver shine

The evening lamps, but, more divine,

The light of love shines over all;

Of love, that says not mine and thine,

But ours, for ours is thine and mine.

They want no guests, to come between

Their tender glances like a screen,

And tell them tales of land and sea,

And whatsoever may betide

The great, forgotten world outside;

They want no guests; they needs must be

Each other's own best company.

III.

THE picture fades; as at a village fair

A showman's views, dissolving into air,

  Again appear transfigured on the screen,

So in my fancy this; and now once more,

In part transfigured, through the open door

  Appears the selfsame scene.

Seated, I see the two again,

But not alone; they entertain

A little angel unaware,

With face as round as is the moon;

A royal guest with flaxen hair,

Who, throned upon his lofty chair,

Drums on the table with his spoon,

Then drops it careless on the floor,

To grasp at things unseen before.

Are these celestial manners? these

`The ways that win, the arts that please?

Ah yes; consider well the guest,

And whatsoe'er he does seems best;

He ruleth by the right divine

Of helplessness, so lately born

In purple chambers of the morn,

As sovereign over thee and thine.

He speaketh not; and yet there lies

A conversation in his eyes;

The golden silence of the Greek,

The gravest wisdom of the wise,

Not spoken in language, but in looks

More legible than printed books,

As if he could but would not speak.

And now, O monarch absolute,

Thy power is put to proof; for, lo!

Resistless, fathomless, and slow,

The nurse comes rustling like the sea,

And pushes back thy chair and thee,

And so good night to King Canute.

IV.

As one who walking in a forest sees

A lovely landscape through the parted trees,

Then sees it not, for boughs that intervene;

Or as we see the moon sometimes revealed

Through drifting clouds, and then again concealed,

So I behold the scene.

There are two guests at table now;

The king, deposed and older grown,

No longer occupies the throne, —

The crown is on his sister's brow;

A Princess from the Fairy Tales,

The very pattern girl of girls,

All covered and embowered in curls,

Rose-tinted from the Isle of Flowers,

And sailing with soft, silken sails

From far-off Dreamland into ours.

Above their bowls with rims of blue

Four azure eyes of deeper hue

Are looking, dreamy with delight;

Limpid as planets that emerge

Above the ocean's rounded. verge,

Soft-shining through the summer night.

Steadfast they gaze, yet nothing see

Beyond the horizon of their bowls;

Nor care they for the world that rolls

With all its freight of troubled souls

Into the days that are to be.

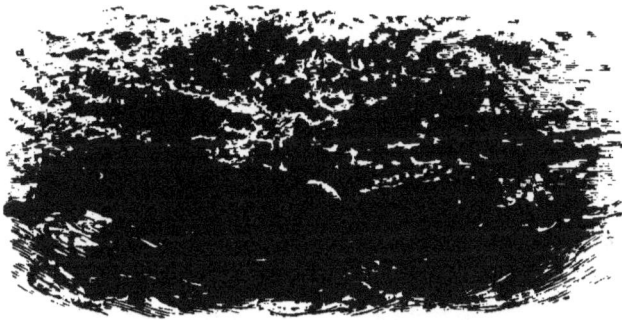

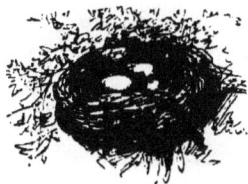

V.

AGAIN the tossing boughs shut out the scene,

Again the drifting vapors intervene,

    And the moon's pallid disk is hidden quite;

And now I see the table wider grown,

As round a pebble into water thrown

      Dilates a ring of light.

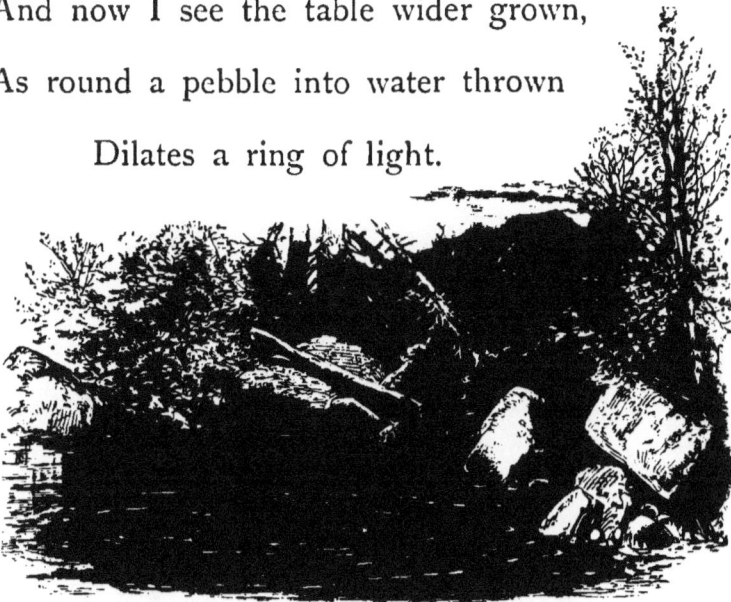

I see the table wider grown,

I see it garlanded with guests,

As if fair Ariadne's Crown

Out of the sky had fallen down;

Maidens within whose tender breasts

A thousand restless hopes and fears,

Forth reaching to the coming years,

Flutter awhile, then quiet lie,

Like timid birds that fain would fly,

But do not dare to leave their nests; —

And youths, who in their strength elate

Challenge the van and front of fate,

Eager as champions to be

In the divine knight-errantry

Of youth, that travels sea and land

Seeking adventures, or pursues,

Through cities, and through solitudes

Frequented by the lyric Muse,

The phantom with the beckoning hand,

That still allures and still eludes.

O sweet illusions of the brain!

O sudden thrills of fire and frost!

The world is bright while ye remain,

And dark and dead when ye are lost!

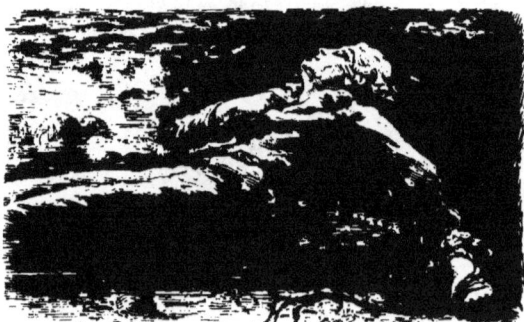

VI.

THE meadow-brook, that seemeth to stand still,

Quickens its current as it nears the mill;

   And so the stream of Time that lingereth

In level places, and so dull appears,

Runs with a swifter current as it nears

   The gloomy mills of Death.

And now, like the magician's scroll,

That in the owner's keeping shrinks

With every wish he speaks or thinks,

Till the last wish consumes the whole,

The table dwindles, and again

I see the two alone remain.

The crown of stars is broken in parts;

Its jewels, brighter than the day,

Have one by one been stolen away

To shine in other homes and hearts.

One is a wanderer now afar

In Ceylon or in Zanzibar,

Or sunny regions of Cathay;

And one is in the boisterous camp

Mid clink of arms and horse's tramp,

And battle's terrible array.

I see the patient mother read,

With aching heart, of wrecks that float

Disabled on those seas remote,

Or of some great heroic deed

On battle-fields, where thousands bleed

To lift one hero into fame.

Anxious she bends her graceful head

Above these chronicles of pain,

And trembles with a secret dread

Lest there among the drowned or slain

She find the one beloved name.

VII.

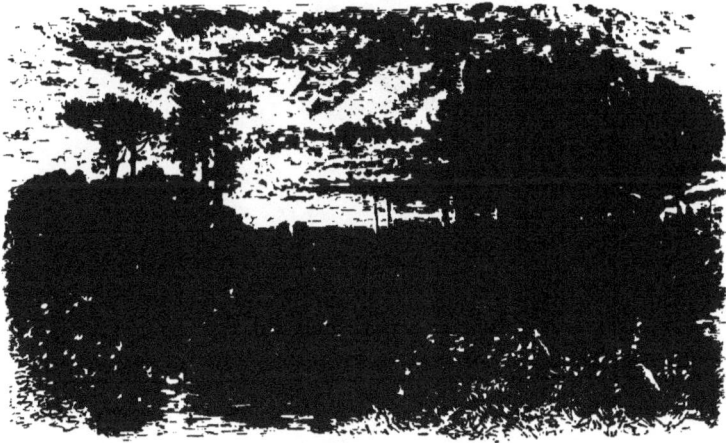

AFTER a day of cloud and wind and rain

Sometimes the setting sun breaks out again,

   And, touching all the darksome woods with light,

Smiles on the fields, until they laugh and sing,

Then like a ruby from the horizon's ring

      Drops down into the night.

What see I now? The night is fair,

The storm of grief, the clouds of care,

The wind, the rain, have passed away;

The lamps are lit, the fires burn bright,

The house is full of life and light:

It is the Golden Wedding day.

The guests come thronging in once more,

Quick footsteps sound along the floor,

The trooping children crowd the stair,

And in and out and everywhere

Flashes along the corridor

The sunshine of their golden hair.

On the round table in the hall

Another Ariadne's Crown

Out of the sky hath fallen down;

More than one Monarch of the Moon

Is drumming with his silver spoon;

The light of love shines over all.

O fortunate, O happy day!

The people sing, the people say.

The ancient bridegroom and the bride,

Serenely smiling on the scene,

Behold, well-pleased, on every side

Their forms and features multiplied,

As the reflection of a light

Between two burnished mirrors gleams,

Or lamps upon a bridge at night

Stretch on and on before the sight,

Till the long vista endless seems.

9 783337 105358